UP CLOSE™

SPACE

PAUL HARRISON

PowerKiDS
press™

New York

Published in 2008 by The Rosen Publishing Group, Inc.
29 East 21st Street, New York, NY 10010

Copyright © 2008 Arcturus Publishing Limited

Author: Paul Harrison
Editor (new edition): Fiona Tulloch
Editor (US editon): Kara Murray
Designer (new edition): Sylvie Rabbe

Picture credits: Corbis: page 3, bottom right; page 6, bottom; page 9, bottom; page 10, top right; page 11, bottom; page 14; NASA: page 3, top; page 5, top right; page 7, top right; Science Photo Library: page 4, top right; page 5, bottom left; page 6, top right; page 8, top and bottom; page 9, top right; page 10, bottom left; page 15, top and bottom.

Library of Congress Cataloging-in-Publication Data

Harrison, Paul, 1969–
  Space / Paul Harrison.
    p. cm. — (Up close)
  Includes index.
  ISBN 978-1-4042-4221-0 (library binding)
  1. Outer space—Juvenile literature. I. Title.
  QB500.22.H37 2008
  520—dc22
                                              2007033484

Manufactured in China

Contents

In the Beginning

Space is all around us and scientists think that it stretches on forever. That can make your head hurt if you think about it for too long. As you'd expect, something that big has a lot of stuff in it. Let's find out where it all came from and what's there.

SOLAR SYSTEM

The term "solar" comes from the Latin word *solaris,* meaning "of the Sun." It's no surprise, then, that our solar system is made up of everything that has the Sun at its center—including planets, comets, and other pieces of space rock. Until recently, we had nine planets in our solar system. Since Pluto was reclassified as a dwarf planet in 2006, we have only eight.

BOOM!

Most scientists believe that the universe began with a bang. Between 10 billion and 15 billion years ago the universe exploded out from a tiny point, hurling out gases that would go on to form the stars and galaxies. This is called the big bang theory and scientists can still pick up an "echo" of radiation from the universe's violent beginning.

Scientists believe that the outside of the Sun is about 10,832° F (6,000° C). But that's nothing compared to the center, which is an astonishing 27,000,032° F (15,000,000° C).

THE BIG ONES

A planet is an object over a certain size that orbits, or travels around, a star. There are three main types of planet: planets that are made mainly from rock, like our own; planets that are made mainly from ice, which are generally the smallest planets; and the biggest planets, which are made from gas. A moon, on the other hand, is an object over a certain size that orbits around a planet rather than a star—just like our Moon.

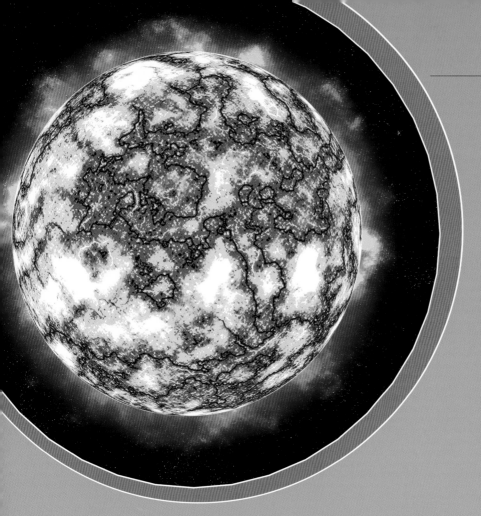

SHINING STAR

Our Sun is just one of the billions of different stars that are out in space. Stars are really just big balls of gas. They come in all sizes and temperatures, from cool dwarf stars, to medium stars like our Sun, to huge, hot giants many, many times the size of our Sun. The only reason our Sun looks different from all the other stars is that we're so much closer to it.

TOO SMALL

Asteroids orbit stars but are too small to be planets even though they are made from the same kind of stuff as the Earthlike planets. Comets are lumps of ice and dust that also orbit a star. Sometimes you can see these for a few days before they're off again on their journey around the Sun. Meteors are small lumps of rock and iron that fall to Earth —you can sometimes see them streaking across the sky.

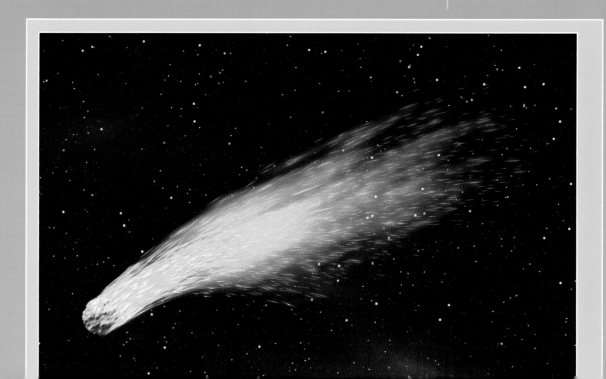

Star Light, Star Bright

The most obvious elements of the night sky are the stars. But these are not just simple points of light—there's much more going on up there than you think.

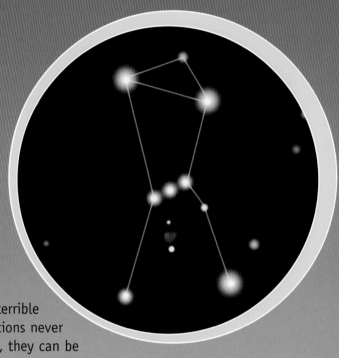

SKY PICTURES

Way back in the mists of time when our ancestors looked at the stars they drew imaginary lines between some of them to make pictures. We call these pictures "constellations." There are 88 different constellations, including Orion the Hunter. Either our ancestors were terrible at drawing, or they were exaggerating, since constellations never look much like what they are supposed to be. However, they can be a handy way of navigating and, in the time before satellite navigation systems or even maps, this was invaluable.

GANGING UP

Although stars seem to be randomly dotted around the night sky, they all belong to groups called galaxies. No one knows how many galaxies there are, but there could well be as many as 200 billion. Our own galaxy is called the Milky Way and contains anywhere between 200 billion and 400 billion stars. It gets its odd name from the milky white line of stars that you can see if you're far away from cities.

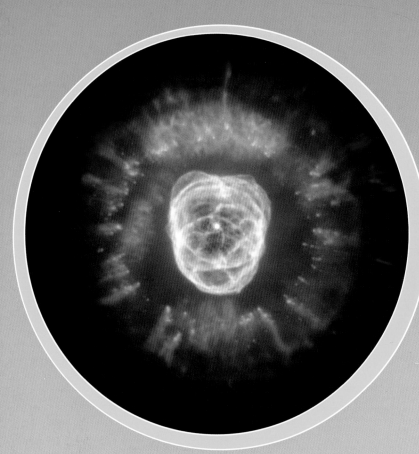

STAR FACTORY

If you have a telescope you may be able to see stars being born. There are huge clouds of gas and dust in space called nebulae and these are the universe's star factories. Within a nebula, smaller clumps of gas and dust sometimes start to spin around each other. As they spin they get hotter and faster and closer together. Eventually they spin out of the nebula and become fully-fledged stars.

Galaxies come in three basic shapes: spiral (shaped like a Catherine wheel), elliptical (oval shaped), and the all-purpose irregular.

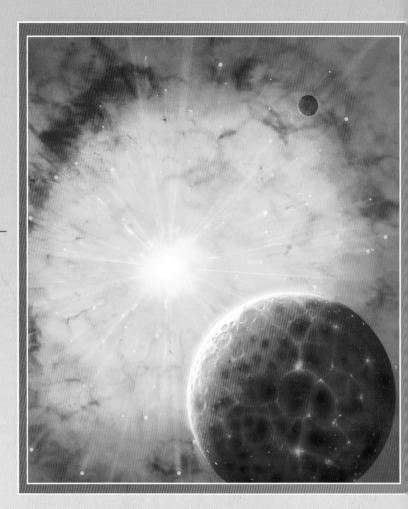

SUPERNOVAS

Stars don't last forever, unfortunately. As a general rule of thumb, the bigger the star the shorter its life will be. However, no matter how big it is, stars always end the same way—in a huge explosion called a supernova! Don't worry, though, our Sun isn't predicted to do this for another five billion years. Although a supernova is the end for one star, it's the beginning for new ones, as the material that's thrown out by the explosion is used in the creation of brand-new stars.

Eyes on the Skies

From the very dawn of the human race, people have found the stars fascinating. As time progressed, so did our ways of looking towards the skies. Our enthusiasm for looking at space has remained undimmed to this day.

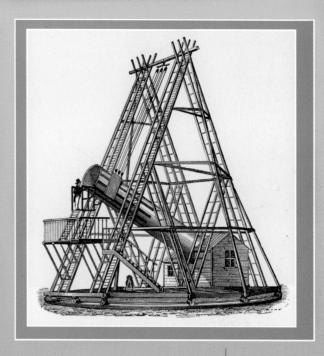

OBSERVE THE OBSERVATORY

The telescopes used by professional space watchers, or astronomers as they are known, are housed in special buildings called observatories. Generally, observatories are found at the tops of hills in sunny climates. That's not so astronomers can get a good view of the countryside while getting a nice tan—it's so they can get a good look at the stars. High up, there is less interference from pollution in the atmosphere and if the country is sunny there are going to be fewer clouds to get in the way.

FIRST GLANCE

Until the invention of the telescope in the early 1600s, people had used their naked eyes to see the stars. That was fine for the time being, but even the most basic of early telescopes was an improvement. The first telescopes allowed people to see the craters on the Moon for the first time and even some of the moons around Jupiter.

HUBBLE BUBBLE

We all know that if you're closer to something you're likely to get a better view of it. Astronomers used that logic when they launched a telescope into space in 1990. The Hubble Telescope, as it is known, is not only closer to the stars; pollution and cloud cover are never going to bother it. Although it might look like a fat drainpipe covered in tinfoil with a couple of solar panels stuck to it, Hubble has sent back pictures from space that the ancients would never have dreamed of, including the birth of stars.

Instead of using one huge telescope, astronomers will often use lots of smaller telescopes pointing at the same bit of the sky.

HEAR! HEAR!

Studying space isn't just about looking at the stars; it involves listening to the universe as well. Objects that are too far away to be seen can sometimes be heard—provided you've got the right equipment. Radio telescopes look a bit like the satellite receiver dishes you see on homes, only much, much bigger. The largest one is the Arecibo Telescope, in Puerto Rico, which is a whopping 1,000 feet (305 m) in diameter. Objects such as stars emit radio waves and this is what radio telescopes, such as the Arecibo, look for.

Infinity and Beyond

I f scientists need to get a closer look at space, they send up a spacecraft. There are many different kinds of spacecraft and they are all designed to do specific jobs.

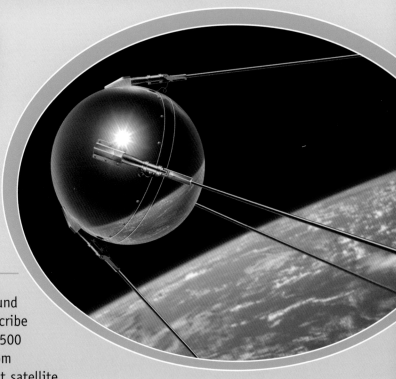

AROUND AND AROUND

Technically, a satellite is any object that orbits around a bigger object, but usually the term is used to describe man-made objects. Currently there are more than 2,500 satellites in space and they do a variety of jobs, from telecommunications to weather forecasting. The first satellite was launched way back in 1957. It was called *Sputnik* and all it did was send out a simple radio signal so that it could be tracked on its orbit around the Earth. It was pretty basic, but it was the first man-made object to orbit the planet.

A CLOSER LOOK

When scientists want to find out about an object in space, the first thing they do is send out a probe. *New Horizons* was launched in January 2006 and was the first probe to orbit around the dwarf planet Pluto to provide scientists with unparalleled information from the edge of our solar system. Some probes, such as *Deep Impact*, aren't designed to orbit an object, but to crash into it. *Deep Impact*—or rather a part of it—was sent to crash into a comet called Tempel 1 so scientists could study what the comet was made of. Understandably, this type of crash is called a hard landing.

OFF FOR GOOD

Some probes are sent from Earth not to orbit anything or land anywhere, but are fired off into space to send back information for as long as their batteries can hold out. The probe that is farthest away from Earth is called *Voyager 1*. It was launched in 1977 and it has travelled about 15 billion miles (24 billion km) from Earth and is getting farther away all the time.

There are more than 8,000 man-made objects floating around in space, from satellites to old bits of spacecraft and things astronauts have dropped.

WHEELY USEFUL

Not all probes that land on objects in space make a hard landing. Others land in a more civilized manner and relay their findings back to Earth. More excitingly, some probes carry rovers with them. Rovers look a bit like remote controlled off-road toys and are designed to drive around the surface of planets. At the moment there are a couple, called *Spirit* and *Opportunity*, tooling around on Mars, and they are sending scientists on Earth some fantastic data about the red planet.

To Boldly Go

The first human being in space was a Russian cosmonaut called Yuri Gagarin. When he travelled around the Earth in 1961, it was just the start of the human adventure in space.

BLAST OFF

The first problem with space exploration is how to get there, which means overcoming Earth's gravity. And to do that you need speed, and lots of it. The only way of getting that kind of speed is by using rockets; the biggest rocket so far was the Saturn V. This giant stood more than 361 feet (110 m) high and was responsible for sending the first American astronauts into space. Despite its huge size it was incredibly fast, travelling at over 5,344 miles per hour (8,600 km/h)!

MOONWALK

Apollo 11 landed on the Moon on July 20, 1969, beginning a new chapter in space exploration. In the words of Neil Armstrong, the first man to walk on the Moon, it was "one small step for man, one giant leap for mankind." The *Apollo 11* team spent two hours on the Moon's surface, collecting samples and taking photographs. Since then, only 11 astronauts have walked on the Moon.

ALL IN THE NAME

More than 400 people have travelled into space, but there are only three countries that have the technology to get people up there: the United States, Russia and China. The United States will take only Americans into space, but the Russians will take anyone who can afford to go. And it all depends on whom you fly with as to what you're called— American and Chinese space travellers are called astronauts, but anyone flying with the Russian space program is called a cosmonaut.

HOME AWAY FROM HOME

Since getting humans into space is such a costly business, it makes sense to get your money's worth by keeping them up there for as long as possible. The best way to do this is to build them a home in orbit around the Earth, or a space station as it is called. As its name suggests, the International Space Station is a joint venture between 16 different countries and is a kind of floating laboratory. At around 240 feet (73 m) at its widest point it's also the biggest man-made object in space.

EXPENSIVE RECYCLING

The space shuttle is the first and so far only reusable spacecraft. It is probably also the most complex machine ever built. It has to be capable of launching itself—with some help from a couple of booster rockets—at over 16,777 miles per hour (27,000 km/h) to get into space and then be able to withstand temperatures of over 1,832° F (1,000 °C) as it reenters the planet's atmosphere. All this comes at a price, though—each shuttle costs around $1.7 billion to build and each mission costs around $450 million!

15

The space shuttle is a space delivery van—its job is to take satellites into space to launch them and to take scientists and supplies to the International Space Station.

TOOLS AT THE READY

As if living aboard the International Space Station wasn't daunting enough, astronauts have to maintain the station, too. How do they do this? By going outside to do it themselves, of course. Astronauts are responsible for everything from connecting cables to protecting the station from debris. These spacewalks are very dangerous due to the orbital velocity of space debris. A grain of sand would travel at the speed of a bullet, puncturing an astronaut's suit and almost certainly causing death.

Mysterious Space

A s you might expect from something as big and as far away as space, there are lots of things that scientists don't know or even understand. Here are some of the big ideas and mysteries.

WHIRLPOOLS IN SPACE

One of the scariest things in space is a black hole. These are incredibly dense parts of space whose gravitational pull works a little like a whirlpool, dragging everything that gets too close into its center. The pull is so strong that even light itself can't escape it. Some scientists believe that black holes occur when giant stars die and that there might even be a hole at the center of our own galaxy!

DOCTOR WHO?

One interesting idea that scientists have is that there might be tunnels in space that lead to different parts of the universe—or even back in time! These tunnels are known as wormholes, and are based on the theory that a person could travel through a black hole and, instead of being crushed at its center, can come out of another hole instead. Scientists know that gravity can affect time (a difficult idea, we know) so theoretically it's possible. We don't have the means to test it yet, but you probably wouldn't like to be the one who tries it first.

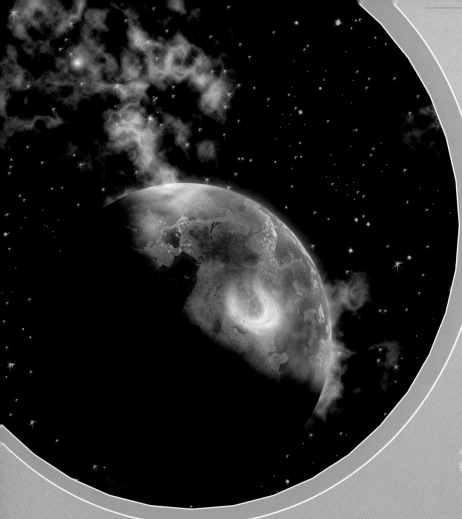

SOMETHING MISSING

Although space has got lots of things in it, such as stars and planets, scientists think that more than 80 percent of space has something in it but they can't tell what it is. It has a gravitational force but there doesn't seem to be anything there. Confused? So are the scientists. They call this missing bit of space "dark matter" as they have no way of detecting what it is as it doesn't give off light or radiation. They guess it must be particles of some sort—they just aren't sure what.

The *Pioneer 10* space probe had a plaque attached to it with drawings of humans, just in case any aliens found it.

WE COME IN PEACE

Are there little green men alive on other planets or are we alone in the universe? Well, many scientists believe that there is life out there—but those beings probably don't go around in flying saucers. Instead they think that they might be tiny, basic microorganisms that can withstand intense heat or cold. Right now, scientists think it's possible that there may be this type of life on Mars, Venus or even on the Moon.

Future Space

Our achievements in space travel, great though they are, are still only the tip of the iceberg. There's so much left to do and explore and there are some really exciting plans in the pipeline.

BIG SLEEP

The biggest problem facing humans trying to explore space is the great distances involved. Even the nearest large galaxy to us is more than 2 million light years away, so even if we could build a spaceship that could travel that far, the human pilots would be dead long before they arrived. The answer seems to be suspended animation—basically slowing down the human body so it doesn't age. It may seem like a far-fetched idea, but scientists have already managed to do it for a short period with mice!

OUT OF THIS WORLD

Space travel used to be the preserve of highly trained pilots, usually from the air force. However, these days anyone can get into space if they've got money—lots of money. In 2001 Dennis Tito became the world's first space tourist when he paid to fly on a Russian rocket and spend some time on the International Space Station. The trip was literally out of this world, and so was the price—it cost a reported $20 million!

DENNIS A. TITO
ДЕННИС А. ТИТО

BACK TO THE MOON

In 1969 Neil Armstrong became the first person to step onto the Moon. Since then, a grand total of only 12 people have visited our nearest neighbor, but things may be about to change; there are plans to build a base on the Moon. The base would work as a laboratory and as a launchpad for other space missions. As the gravity on the Moon is six times weaker than on Earth, it would take a lot less fuel to reach escape velocity, making space exploration much cheaper.

A light year is the distance something would travel if it moved at the speed of light for a year.

OFF TO MARS?

During the 1960s there was a "space race" between America and the Soviet Union and the winner was the first one to get to the Moon. Now a new space race is beginning and the objective is to get a person to Mars. Before that can happen, though, a number of probes need to land on the planet to tell us what the conditions there are really like. After all, it would be a bit of a shame to travel all that way only to have something go wrong the minute an astronaut steps out of his rocket!

Victor Habbick

DIZZY HEIGHTS

Following Arthur C. Clarke's novel
Fountains of Paradise, NASA is planning
to build a huge 24,855 mile (40,000 km)
high elevator reaching from Earth into
space. Its base—taller than the Eiffel
Tower—would have a cable attached to
the top with the other end of the cable
being anchored to a space station, or
even an asteroid! People would be
transported into space by trains running
on tough fibers called carbon nanotubes.
Although they are as strong as diamonds,
the cables are no thicker than a human
hair. Feeling brave? You'll have to wait
until the end of this century to try it out.

AFTER HUBBLE

The Hubble Telescope may be the
latest technology, but what's
next? Why, the Next Generation
Space Telescope, of course.
Although it sounds like something
from *Star Trek*, this telescope uses
infrared technology; the same
thing we use in a TV remote
control! The telescope is
scheduled to be launched in
2010. It will have a huge mirror,
which it will use to study the
formation of stars and the history
of the universe. By using infrared
wavelengths, the telescope will
also be able to study dark matter,
which has been puzzling scientists
for generations.

Glossary

Asteroid (AS-teh-royd)
One of many rocky objects, varying in size that circle the Sun.

Astronomer (uh-STRAH-nuh-mer)
Someone who studies the universe and objects that exist naturally in space, such as the Moon, the Sun, planets, and stars.

Galaxy (GA-lik-see)
A very large group of stars in the universe.

Gravity (GRA-vih-tee)
The force that pulls things to the ground.

Light year (LYT YIR)
The distance light travels in one year.

Microorganism (my-kroh-OR-guh-nih-zum)
A living thing that is too small to be seen unless through a microscope.

Nebula (NEH-byuh-luh)
A cloud of gas or dust in space.

Orbit (OR-bit)
To follow a curved path around a planet or star.

Probe (PROHB)
A small spacecraft, with no one travelling in it, sent into space to make measurements and send back information to scientists on Earth.

Rover (ROH-vur)
A vehicle used to explore the surface of a planet.

Satellite (SA-tih-lyt)
An object moving around a larger object in space.

Solar system (SOH-ler SIS-tem)
The Sun and the eight planets that move around it, including Earth.

Supernova (soo-per-NOH-vuh)
A star that has exploded and glows very brightly.

Universe (YOO-nih-vers)
Everything that exists, including all the stars, planets, and galaxies, in space.

Velocity (veh-LO-suh-tee)
The speed at which an object travels.

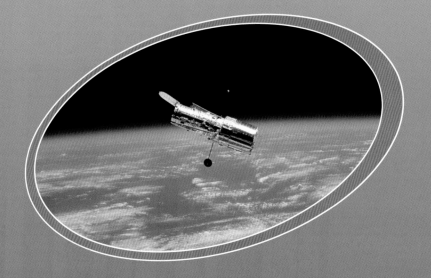

Further Reading

Oxford First Book of Space (Oxford First Series)
by Andrew Langley
New York: Oxford University Press, 2003

Moon Landing (Speedy Reads)
by Nick Arnold
New York: Scholastic, 2001

Space (Kingfisher Voyages)
by Mike Goldsmith
Boston: Kingfisher Books Ltd, 2006

The Usborne First Encyclopedia of Space
by Paul Dowswell
London, UK: Usborne Publishing Ltd, 2001

Living on Other Worlds (Our Universe)
by Gregory L. Vogt
Chicago: Raintree, 2000

WEB SITES
Due to the changing nature of Internet links, PowerKids Press has developed an online list of Web sites related to the subject of this book. This site is updated regularly. Please use this link to access the list:
www.powerkidslinks.com/upcl/space/

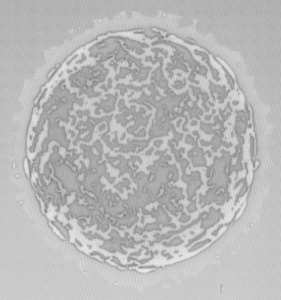

Index